EXCITING OUTDOOR CAREERS

# VOLCANOLOGIST

DWAYNE HICKS

PowerKiDS press

Published in 2026 by The Rosen Publishing Group, Inc.
2544 Clinton Street, Buffalo, NY 14224



First Edition

Editor: Greg Roza
Book Design: Rachel Rising

Photo Credits: Cover, Mystockimages/iStock.com; Cover, pp. 1, 3–24 Iurii Vlasenko/Shutterstock.com; Cover, pp. 1, 3–24 klyaksun/Shutterstock.com; p. 5 fogcatcher/Shutterstock.com; p. 7 Jakub Korczyk/Shutterstock.com; p. 8 Drazen Zigic/Shutterstock.com; p. 9 Atthapon Niyom/Shutterstock.com; p. 10 ImageBank4u/Shutterstock.com; p. 11 https://commons.wikimedia.org/wiki/File:Sampling_lava_with_hammer_and_bucket.jpg; p. 13 smile plus/Shutterstock.com; p. 13 Chokniti-Studio/Shutterstock.com; p. 14 Gianfranco Vivi/Shutterstock.com; p. 15 Nava Fedaeff/Shutterstock.com; p. 17 https://commons.wikimedia.org/wiki/File:Installation-gps-mast-station-brsp-mount-hood.jpg; p. 18 https://commons.wikimedia.org/wiki/File:MSH80_david_johnston_at_camp_05-17-80_med.jpg; p. 19 https://commons.wikimedia.org/wiki/File:KatiaMauriceKrafft.jpg; p. 20 Evgeny Haritonov/Shutterstock.com; p. 21 Rido/Shutterstock.com.



Cataloging-in-Publication Data

Names: Hicks, Dwayne.
Title: Volcanologist / Dwayne Hicks.
Description: Buffalo, New York : PowerKids Press, 2026. | Series: Exciting outdoor careers | Includes glossary and index.
Identifiers: ISBN 9781499457278 (pbk.) | ISBN 9781499457285 (library bound) | ISBN 9781499457292 (ebook)
Subjects: LCSH: Volcanologists–Juvenile literature. | Volcanologists–Vocational guidance–Juvenile literature.
Classification: LCC QE521.3 H54 2026 | DDC 551.21–dc23

Manufactured in China

CPSIA Compliance Information: Batch #QWPK26. For Further Information contact Rosen Publishing at 1-800-237-9932.

# CONTENTS


AN EXPLOSIVE CAREER! .......... 4
KINDS OF VOLCANOLOGISTS .......... 6
EDUCATION AND TRAINING .......... 8
IN THE FIELD .......... 10
IN THE LAB .......... 12
EDUCATING THE PUBLIC .......... 14
PROTECTING THE PUBLIC .......... 16
FAMOUS VOLCANOLOGISTS .......... 18
THE RIGHT CAREER FOR ME? .......... 20
GLOSSARY .......... 22
FOR MORE INFORMATION .......... 23
INDEX .......... 24

# AN EXPLOSIVE CAREER!

Volcanologists are scientists who work in extremely dangerous locations. Although a volcanologist's job often includes laboratory work, research, and report writing, it also includes studying volcanic activity, mapping volcanic formations, and collecting rock samples.

Volcanology is a branch of geology, which is the study of Earth, its structure, and its features. Geologists reveal Earth's history and processes through a careful study of its rocks, **fossils**, and landforms. Of all the kinds of geologists, volcanologists are the rock stars of geology! They put themselves in real danger to learn more about Earth and its ground-shaking forces.

## OUTDOOR ADVICE

Pliny the Elder—a Roman writer and scientist—died in 79 CE in a historic **volcanic eruption**. His son, Pliny the Younger, described the **earthquakes** before the eruption, the eruption itself, the column of ash, **pyroclastic flows**, and even a **tsunami**. Many consider him the first volcanologist.

Do you have what it takes to be a volcanologist? Let's find out!

# KINDS OF VOLCANOLOGISTS

Volcanologists may specialize in a specific area. Volcanic geologists study the formation and structure of volcanoes. Seismologists study the seismic activity—earthquakes—around volcanoes. Geochemists study the chemical make-up of volcanic rocks and gases. Hazard scientists study volcanic eruptions to asses the danger they pose to the surrounding wildlife, land, people, and property.

Not all volcanologists get up close and personal with volcanoes. Remote sensing **technicians** use **satellites** and drones to **monitor** volcanoes. Ecologists study the affects volcanoes have on the natural world. Many volcanologists go on to become teachers, while others work with governments to educate the public on the dangers of volcanoes.

**OUTDOOR ADVICE**

Some scientists study volcanoes on other planets!

Volcanologists usually don't work alone. They work with other volcanologists and scientists, as well as government officials and emergency workers.

# EDUCATION AND TRAINING

Many volcanologists have a PhD. This is a college degree that can take eight or more years to complete. This includes classroom study, lab work, and fieldwork. At the least, volcanologists usually need a four-year degree in one of several fields, including physics, chemistry, or earth science.

Aside from a college education in science, there are other ways to prepare for a career as a volcanologist. Since fieldwork includes hiking through wilderness and up the sides of volcanoes, volcanologists need to be physically fit. You can work on that now by joining a sport, riding a bike, running, or swimming.

## OUTDOOR ADVICE

Volcanology can be dangerous. Volcanologists should be trained in first aid and survival skills.

Once samples have been collected from the area around a volcano, they are transported to a labratory for study. College students are sometimes invoved with laboratry work.

# IN THE FIELD

Volcanologists often risk their lives while in the field, especially when around an active volcano! There are numerous important reasons for doing fieldwork. Like many scientists, volcanologists observe volcanic sites and collect information. They examine volcanic rocks, and even collect melted rock called lava, which has recently flowed out of the earth. These observations and samples are often later studied in a lab.

### OUTDOOR ADVICE

Modern volcanologists can use drones to monitor and even take samples from active volcanoes. This keeps workers safer.

Volcanologists monitor, or keep an eye on, volcanic areas. They watch for volcanic activity, such as lava flows and eruptions. They also watch for changes in the land that might tell them a volcano is going to erupt soon.

Volcanologists often need to hike and climb over rough land to reach remote locations in order to take samples.

# IN THE LAB

Although laboratory work isn't as exciting as fieldwork—or as dangerous—it's still important to a volcanologist's job. Volcanologists and other scientists study information that has been gathered in the field. They do experiments on samples to find out what they're made of. Samples can include rocks, hardened lava, volcanic gases, and sometimes plants and animals. This work gives us a better understanding of volcanoes.

Laboratories are filled with high-tech tools to help volcanologists do their jobs. Computers help volcanologists identify patterns in the information they've collected. They also help volcanologists guess when a volcano may erupt again.

## OUTDOOR ADVICE

Geologists and volcanologists sometimes take core samples from the soil and rock around a volcano to study in a lab. These are long, tubelike sections of dirt removed from the ground.

This volcanologist is running tests on several rock samples taken from the area around an active volcano.

# EDUCATING THE PUBLIC

Some volcanologists choose to become educators and share their knowledge with others. This often means becoming a high school or college teacher. This is a great chance for volcanologists to share their knowledge with other budding geologists. Some volcanologists work in research centers where they can guide less experienced volcanologists. Still others work for national parks, observatories, and natural museums where they guide, entertain, and educate the public.

Like other scientists, many volcanologists write reports on their findings and publish them in science journals or present them at conferences. This helps other volcanologists, but also helps educate the public on the dangers of volcanic activity.

Some volcanologists are hired to lead tours of people who want to view active volcanoes. They teach people on the tour all about the volcano and it's natural history. They also keep tourists safe.

# PROTECTING THE PUBLIC

The work that volcanologists do is essential in keeping the public safe from volcano-related events. Their work is used to **mitigate** the dangers that go along with volcanoes and volcanic eruptions. They create plans for responding to volcanic activity.

The U.S. Geological Survey (USGS) is a government group that hires volcanologists to monitor and asses volcanic dangers. The USGS set up field stations that record seismic and volcanic activity. Volcanologists use this information to make **predictions** about volcanic eruptions. This helps keep people safe. They also work with local communities to raise public awareness of the dangers posed by volcanoes.

## OUTDOOR ADVICE

The United State has about 168 active volcanoes. Most are found in Alaska and along the Pacific Northwest coast. The Hawaiian Islands are volcanic islands.

These volcanologists work for the USGS. They are setting up a volcanic activity field station to monitor activity around Mount Hood in Oregon.

# FAMOUS VOLCANOLOGISTS

David A. Johnston was an incredibly brave and talented volcanologist who worked for the USGS. In May 1980, Johnston was in the field monitoring the increasing seismic activity around Mount St. Helens in Washington State. The sudden eruption of this volcano on May 18 killed 57 people, including Johnston.

Today, several locations in Mount St. Helens National Volcanic Monument are named after Johnston, including the Johnston Ridge and the Johnston Ridge **Observatory**.

Katia and Maurice Krafft were French scientists who studied and filmed volcanoes as a couple for 20 years. They recorded eruptions, studied them, and explained dangers to the public. The Kraffts died in a pyroclastic flow while filming the eruption of Mount Unzen in Japan on June 3, 1991.

# THE RIGHT CAREER FOR ME?

Volcanology isn't a career for everyone. It takes years of college training, and the job itself is a dangerous one. However, it's a good paying job. The average pay for volcanologists is about $100,000. If you work your way up to a leadership position, you can make $200,000 or more. Also, volcanology is a growing career, meaning the number of jobs available is on the rise.

Do you think a career as a volcanologist is a good fit for you? It's a highly dangerous job that requires long-term training. It's not an easy decision to make! Good thing you have some time to think it over.

You don't need to get up close to a volcano to study them. You may even build a volcano model that really erupts in science class!

# GLOSSARY

**earthquake:** A sudden and violent shaking of the ground caused by movements within Earth's crust.

**fossil:** A trace, print, or the remains of a plant or animal from long ago that is preserved in earth or rock.

**mitigate:** To cause something to be less dangerous or harsh.

**monitor:** To observe and check the progress of something.

**observatory:** A building or other structure that provides a wide view of the surroundings.

**prediction:** A guess made as to what will happen in the future based on facts or knowledge.

**pyroclastic flow:** An extremely hot mixture of rock, dust, and gas from a volcano that moves at very high speed away from the volcano when it erupts.

**satellite:** A spacecraft placed in orbit around Earth, a moon, or a planet to collect information or for communication.

**technician:** A specialist in the technical or mechanical details of a subject or career.

**tsunami:** A great sea wave most often produced by an earthquake or volcanic eruption under the sea.

**volcanic eruption:** An event in which a volcano, or an opening in a planet's surface through which hot, liquid rock sometimes flows, sends out that rock and other substances in a sudden explosion.

# FOR MORE INFORMATION

## BOOKS

Borgert-Spaniol, Megan. *STEM Underground.* Minneapolis, MN: ABDO, 2023.

Lake, Theia. *Volcanologist.* Buffalo, NY: Gareth Stevens, 2024.

## WEBSITES

**He Died as He Lived: David A. Johnston**
*unapologeticnerd.com/2015/05/18/he-died-as-he-lived-david-a-johnston*
Read about the life and work of David Johnston and the contributions he made to volcanology.

**Scientist Katia Krafft Stood on the Edge of Active Volcano Craters**
*www.atlasobscura.com/articles/katia-krafft-volcanologist*
Learn more about the amazing life and work of Katia and Maurice Krafft.

# INDEX


C

core samples, 12, 13

E

earthquakes, 4, 6
educating the public, 6, 14, 15, 19
eruption, 4, 6, 10, 12, 16, 18, 19, 21

F

field stations, 16, 17
fieldwork, 8, 10, 12, 18

J

Johnston, David A., 18

K

Krafft, Katia and Maurice, 19

L

laboratory, 4, 8, 9, 10, 12, 13
lava, 10, 12

P

protecting the public, 16
pyroclastic flows, 4, 19

S

samples, 4, 9, 10, 11, 12, 13

U

U.S. Geological Survey, 16, 17, 18